Animal Behavior Investigations

# ANIMAL ABILITIES

By Tracie Santos

Rourke Educational Media
A Division of Carson Dellosa Education

ROURKE'S
# SCHOOL to HOME
CONNECTIONS

# BEFORE AND DURING READING ACTIVITIES

## Before Reading: *Building Background Knowledge and Vocabulary*

Building background knowledge can help children process new information and build upon what they already know. Before reading a book, it is important to tap into what children already know about the topic. This will help them develop their vocabulary and increase their reading comprehension.

### Questions and Activities to Build Background Knowledge:

1. Look at the front cover of the book and read the title. What do you think this book will be about?
2. What do you already know about this topic?
3. Take a book walk and skim the pages. Look at the table of contents, photographs, captions, and bold words. Did these text features give you any information or predictions about what you will read in this book?

### Vocabulary: *Vocabulary Is Key to Reading Comprehension*

Use the following directions to prompt a conversation about each word.

- Read the vocabulary words.
- What comes to mind when you see each word?
- What do you think each word means?

**Vocabulary Words:**

- *abilities*
- *camouflage*
- *echolocation*
- *enclosure*
- *magnetic fields*
- *mimic*
- *predators*
- *tuberculosis*

## During Reading: *Reading for Meaning and Understanding*

To achieve deep comprehension of a book, children are encouraged to use close reading strategies. During reading, it is important to have children stop and make connections. These connections result in deeper analysis and understanding of a book.

### Close Reading a Text

During reading, have children stop and talk about the following:

- Any confusing parts
- Any unknown words
- Text to text, text to self, text to world connections
- The main idea in each chapter or heading

Encourage children to use context clues to determine the meaning of any unknown words. These strategies will help children learn to analyze the text more thoroughly as they read.

When you are finished reading this book, turn to the next-to-last page for **Text-Dependent Questions** and an **Extension Activity**.

# Table of Contents

# Case Study: Car Alarms at the Zoo

There is a mystery at the zoo. The staff have been hearing strange things near the superb lyrebird **enclosure**. It sounds like people talking. As the staff get closer, they hear the sound of cameras clicking and car alarms!

The zookeeper studies the area. She visits the birds' enclosure at the same time of day as the people hearing the noises. She hears the noises, too!

Superb Lyrebird Behavior
Superb Lyrebird

No cars are allowed in that part of the zoo. There were no other people around when the noises were heard. This is the only part of the zoo where the zookeeper hears these sounds. Why are these sounds coming from inside the cage? As you learn more about animal **abilities**, you can help the zookeeper solve this mystery.

# Special Talents

The animal world is full of incredible abilities. These are like an animal's superpower. Sometimes they enable animals to use other senses, such as hearing, in a new way. Other abilities are unique to a kind or group of animals.

Loggerhead Sea Turtle

African Giant Pouched Rat

Orca

Mimic Octopus

Their abilities help animals survive by making it easier for them to find food, mates, or shelter. They can also protect animals against **predators**. These abilities can change how animals interact with the world around them.

Big Brown Bat

Vampire Bats

Dolphins

# Abilities in Action

Some animals have special senses. For example, most bats, dolphins, and some whales can send out sound waves. These waves hit objects and bounce back. This helps the animals avoid the objects and find food. This sense is called **echolocation**.

Abilities can even help animals find their way over long distances. Loggerhead sea turtles swim thousands of miles each year to find food and build nests. They sense the Earth's **magnetic fields** and use them to make sure they are swimming in the right direction.

Common Bent-Wing Bat

## A Way Home

Adult loggerhead sea turtles lay their eggs on beaches and swim away right after. The young turtles hatch and swim to areas with more food. Many years later, they use magnetic fields to return to the beaches where they hatched. Then, they lay their eggs there.

Other abilities make animals sound or look like other things. Some birds can **mimic** the songs of different birds. They do this to communicate with birds around them, including mates.

Common Octopus

Big Blue Octopus

Many kinds of octopuses use **camouflage** to make themselves look like sand, rocks, or other things around them. This makes it easier for them to hide from predators. It also lets them sneak up on other animals while hunting.

A few animals have special abilities that can help humans. African giant pouched rats can smell scents that other animals can't, including the smell of explosive land mines. The rats can be trained to run through a field and alert humans when they smell a buried land mine. The rats don't weigh enough to set off the buried land mines. This way, the mines can be safely removed.

## Magawa the Hero Rat

A group named APOPO trains the rats to find mines and alert humans. One rat is so good at finding land mines that he has been given a gold medal. Magawa has found almost 40 land mines, more than any other African giant pouched rat.

These giant rats can save lives in another way, too. They are trained to smell samples of a person's saliva. The rats can tell whether a person has **tuberculosis**, a deadly disease. This "rat test" is much faster than other kinds of testing.

# What's That Sound?

The zookeeper studies the lyrebirds. Now, she has the answer. In the wild, the birds mimic the sounds they hear around them. These are usually the songs of other birds. But the lyrebirds in the zoo hear different sounds than birds in the wild. Do you know what's happening?

The lyrebirds at the zoo are mimicking the sounds that people make around them. That's why they make noises such as car alarms and cameras. The mystery is solved!

To learn more about the case study and scientists in this book, search online for "Chook the lyrebird" and "Adelaide Zoo."

# Get Creative! Animal Extension Activity

The examples in this book are just a few of the incredible abilities in the animal world. How could you study one that was not described here? Research other animal abilities and choose one. Design a study that could help you learn more about it.

When designing your study, think about the following things:

1. What animal ability are you studying?
2. What information about the ability would you like to find?
3. How do you think the ability helps the animal?
4. How could you test whether your guess was correct?

**Design your study. Make sure it is safe for living things. Write a plan for your study and share it with a friend.**

# Glossary

**abilities** (uh-BIL-i-teez): skills

**camouflage** (KAM-uh-flahzh): a disguise or a natural coloring that allows animals, people, or objects to hide by making them look like their surroundings

**echolocation** (EK-oh loh-KAY-shun): the ability to sense objects by sending out sound waves and hearing where they bounce back

**enclosure** (en-KLOH-zhur): an area surrounded by a barrier, such as a fence or walls

**magnetic fields** (mag-NET-ik feeldz): the areas around a magnetic pole or electric current that have the power to attract metals

**mimic** (MIM-ik): to imitate someone or something else

**predators** (PRED-uh-turs): animals that live by hunting other animals for food

**tuberculosis** (tu-bur-kyuh-LOH-sis): a highly contagious disease caused by bacteria that usually affects the lungs

# Index



# Text-Dependent Questions

1. What are four abilities in this book that help animals survive? How do they do that?
2. Why did the superb lyrebirds at the zoo make such strange sounds?
3. Why are trained African giant pouched rats used to detect land mines?
4. What ability do loggerhead sea turtles use to navigate?

# Further Reading


Grunbaum, Mara. *Sea Turtles*, Children's Press, 2018

Koch, Falynn. *Science Comics: Bats*, First Second Books, 2017

Montgomery, Sy. *The Octopus Scientists*, HMH, 2015.

# About the Author

**Tracie Santos has loved animals since she was very young. She has worked in zoos and aquariums and studied many amazing animal behaviors. She lives in Columbus, Ohio, with her two hairless cats, whose special ability is always knowing when it is time for breakfast.**



www.rourkeeducationalmedia.com

PHOTO CREDIT: Cover ©Alex C Maisey, ©CraigRJD; back cover ©RayK Photos; p5 ©Nathan King, ©Susan Flashman; p6 ©PA Images; p7 ©Dave Watts; p8 ©tracielouise, hocus-focus; p9 ©APOPO, ©Nick Grobler, ©Trueog; p10 ©karen crewe, ©Danny Ye; p11 ©imageBROKER, ©Corina Daniela Obertas, ©32 Pixels; p12 ©JAH; p13 ©Shane Myers Photography, ©Florida Chuck, ©32 Pixels; p14 ©Kittichai numa, ©32 Pixels; p15 ©JodiJacobson, ©Placebo365; p16A ©Xavier Rossi, ©32 Pixels; p16 ©32 Pixels; p16b ©APOPO; p17 ©CB2/ZOB, ©32 Pixels; p18 ©David Noble, p19 ©tracielouise; p20 ©Natee Photo; p21 ©BonD80, ©Boltenkoff, ©photo, ©Hurst Photo, ©Frame Art; p24 ©Taryn Lindsey

Edited by: Tracie Santos
Cover and interior layout by: Tammy Ortner

**Library of Congress PCN Data**

Animal Abilities / Tracie Santos
(Animal Behavior Investigations)
ISBN 978-1-73164-889-1 (hard cover)(alk. paper)
ISBN 978-1-73164-837-2 (soft cover)
ISBN 9-781-73164-941-6 (e-Book)
ISBN 978-1-73164-993-5 (ePub)
Library of Congress Control Number: 2021935272

Rourke Educational Media
Printed in the United States of America
06-0202511937